The Real Nature of Time:
An Analysis of Physics, Prophecy, and Time Travel
Experiences

Copyright Page

This book is copyrighted for 2020

The Real Nature of Time:
An Analysis of Physics, Prophecy, and Time
Travel Experiences

The Crazy and Out of the Box Series Book 7

By Martin K. Ettington

ISBN: 9798665543420

Printed in the United States of America

I0480568

The Real Nature of Time:
An Analysis of Physics, Prophecy, and Time Travel Experiences

The Real Nature of Time:
An Analysis of Physics, Prophecy, and Time Travel Experiences

The real nature of time is still a mystery after thousands of years of philosophers and scientists analyzing it and generating theories about time.

Prophecy has existed for most of humanity's time on Earth and there is a lot of reason to think that seeing the future does see some of what happens. I also wrote two books on this subject including the history of prophecy and my own experiences. The titles include "Prophecy: A History and How to Guide" and "Use Intuition and Prophecy to Improve Your Life-By an Adept"

In the twentieth century Albert Einstein gave the world his Theories of Special Relativity and General Relativity. He gave us a new way to view time in the world. The result of his analysis of time, space, and gravity has affected our scientific understanding about the world and Universe we live in. This interests me because I studied Physics before switching to Engineering in my Junior Year at University.

Then there are people who claim to have had time travel experiences. I found enough of these stories for my book titled "Real Time Travel Stories from a Psychic Engineer".

With all that I've learned from my studies of Physics, my prophecy experiences, and research on real time travel stories this is my effort to try to integrate all of this information together.

I hope you will find this book interesting and it will provoke your thinking.

The Real Nature of Time:
An Analysis of Physics, Prophecy, and Time Travel
Experiences

The Real Nature of Time:
An Analysis of Physics, Prophecy, and Time Travel Experiences

Other books by Martin K. Ettington

Spiritual and Metaphysics Books:
Prophecy: A History and How to Guide
God Like Powers and Abilities
Enlightenment for Newbies
Removing Illusions to Find True Happiness
Using the Scientific Method to Study the
          Paranormal
A Compendium of Metaphysics and How
          to Guides (Six books together
          in one volume)
Love from the Heart
The Enlightenment Experience
Learn Your Soul's Purpose
Pursuing Enlightenment
A Modern Man's Search for Truth
Use Intuition and Prophecy to Improve
          Your Life
The Handbook of Spiritual and Energy
          Healing

Longevity & Immortality:
Physical Immortality: A History and How
          to Guide
The Commentaries of Living Immortals
Records of Extremely Long Lived Persons
Enlightenment and Immortality
Longevity Improvements from Science
The 10 Principles of Personal Longevity
Telomeres & Longevity
The Diets and Lifestyles of the Worlds
          Oldest Peoples
The Longevity Six Books Bundle

Science Fiction:
Out of This Universe
Personal Freedom-Parts 1 & 2
The Psychic Soldier Series:
    Book 1-Himalayan Journey
    Book 2-A Soldier is Born
    Book 3-Fighting For Right
    Book 4-Earth Protector
The Immortality Sci Fi Bundle

The God Like Powers Series:
Human Invisibility
Invulnerability and Shielding
Teleportation
Psychokinesis
Our Energy Body, Auras, and
Thoughtforms

The God Like Powers Series—
          Volume 1 Compilation
The Yoga Discovery Series:
Yoga-An Ancient Art Form
Hatha Yoga-Helping you Live Better
Raja Yoga-Through the Ages
The Yoga Discovery Package

Business & Coaching Books:
Creating, Publishing, & Marketing
          Practitioner Ebooks
Building a Successful Longevity
          Coaching Business
Why Become a Coach?
The Professional Coaching Success Trilogy
2020-Make Money Writing and Selling
          Books
The 2020 Handbook of High Paying Work
          Without a College Degree

Science, Technology, and Misc.
Future Predictions By and Engineer & Seer
The Unusual Science & Technology Bundle
The Real Atlantis-In the Eye of the Sahara
Are Cryptozoological Animals Real or
          Imaginary?
Real Time Travel Stories From a Psychic
          Engineer
Removing Limits On Our Consciousness-
          And Thinking Outside the Box
33 Incredible True Survival Stories
How to Survive Anything: From the
          Wilderness to Man Made
          Disasters
All About Mars Journeys and Settlement
Mining the Asteroid Belt

Ancient History
The Real Atlantis-In the Eye of the Sahara
Ancient & Prehistoric Civilizations
Ancient & Prehistoric Civilizations-Book Two
The History of Antediluvian Giants
The Antediluvian History of Earth
Ancient Underground Cities and Tunnels
Strange Objects Which Should Not Exist
Strange and Ancient Places in the USA
A Theory of Ancient Prehistory And Giant
          Aliens
Aliens and Space
Aliens and Secret Technology
Aliens Are Already Among Us

The Real Nature of Time:
An Analysis of Physics, Prophecy, and Time Travel Experiences

These books are all available in digital and printed formats from my website and on Amazon, Barnes & Noble, Apple ITunes, and many other sites

My Books Website is: http://mkettingtonbooks.com

Signup for our Mailing List to get the following:

1) A discount coupon for 25% discount on all books on our site

2) Occasional Notices of new books available

3) Occasional Email on other offerings of ours (Monthly)

Go to this link to sign-up:

http://personal-longevity.com/mkebooks/emailsignup/

And click this link to get the FREE 102 page Ebook titled "Secrets of Many Things"

If you have any questions about this book or other subjects please contact the Author at:

mke@mkettingtonbooks.com

The Real Nature of Time:
An Analysis of Physics, Prophecy, and Time Travel
Experiences

The Real Nature of Time:
An Analysis of Physics, Prophecy, and Time Travel
Experiences

Table of Contents

The Real Nature of Time:
An Analysis of Physics, Prophecy, and Time Travel
Experiences

The Real Nature of Time:
An Analysis of Physics, Prophecy, and Time Travel
Experiences

1.0 Introduction

The real nature of time is still a mystery after thousands of years of philosophers and scientists analyzing it and theories about it. My hope for this book is to use the knowledge I've collected on different aspects of time to see how it all connects together. I also encourage you readers to review my work and come up with your own ideas.

Prophecy has existed for most of humanity's time on Earth and there is a lot of reason to think that seeing the future does see some of what happens. I also wrote two books on this subject including the history of prophecy and my own experiences. The titles include "Prophecy: A History and How to Guide" and "Use Intuition and Prophecy to Improve Your Life-By an Adept"

In the twentieth century Albert Einstein gave the world his Theories of Special Relativity and General Relativity. He gave us a new way to view time in the world. The result of his analysis of time, space, and gravity has affected our scientific understanding about the world and Universe we live in. This interests me because I studied Physics before switching to Engineering in my Junior Year at University.

Then there are people who claim to have had time travel experiences. I found enough of these stories for my book titled "Real Time Travel Stories from a Psychic Engineer".

With all that I've learned from my studies of Physics, my prophecy experiences, and my research on real time travel stories this is my effort to try to integrate all of this information together.

The Real Nature of Time:
An Analysis of Physics, Prophecy, and Time Travel
Experiences

2.0 Relativistic Space and Time

Albert Einstein

The idea of relativistic time is a direct result of Albert Einstein's Theory of Relativity

Since Albert Einstein published his Theory of Relativity (the Special Theory in 1905, and the General Theory in 1916), our understanding of time has changed dramatically, and the traditional Newtonian idea of absolute time and space has been superseded by the notion of time as one dimension of space-time in special relativity, and of dynamically curved space-time in general relativity.

It was Einstein's genius to realize that the speed of light is absolute, invariable and cannot be exceeded (and indeed that the speed of light is actually more fundamental than either time or space). In relativity, time is certainly an integral part of the very fabric of the universe and cannot exist apart from the universe, but, if the speed of light is invariable and absolute, Einstein realized, both space and time must be flexible and relative to accommodate this.

Although much of Einstein's work is often considered "difficult" or "counter-intuitive", his theories have proved (both in laboratory experiments and in astronomical observations) to be a remarkably accurate model of reality, indeed much more accurate than Newtonian physics, and applicable in a much wider range of circumstances and conditions.

Space-Time

One aspect of Einstein's Special Theory of Relativity is that we now understand that space and time are merged inextricably into four-dimensional space-time, rather than the three dimensions of space and a totally separate time dimension envisaged by Descartes in the 17th Century and taken for granted by all classical physicists after him. With this insight, time effectively becomes just part of a coordinate specifying an object's position in space-time.

It was Hermann Minkowski, Einstein's one-time teacher and colleague, who gave us the classic interpretation of Einstein's Special Theory of Relativity. Minkowski introduced the relativity concept of proper time, the actual elapsed time between two events as measured by a clock that passes through both events. Proper time therefore depends not only on the events themselves but also on the motion of the clock between the events. By contrast, what Minkowski called coordinate time is the apparent time between two events as measured by a distant observer using that observer's own method of assigning a time to an event.

An event is both a place and a time, and can be represented by a particular point in space-time, i.e. a point in space at a particular moment in time. Space-time as a whole can therefore be thought of as a collection of an infinite number of events. The complete history of a particular point in space is represented by a line in space-time (known as a world line), and the past, present and future accessible to a particular object at a particular time can be represented by a three dimensional light cone (or Minkowski space-time diagram), which is defined by the limiting value of the speed of light, which intersects at the here-and-now, and through which the object's world line runs its course.

Modern physicists therefore do not regard time as "passing" or "flowing" in the old-fashioned sense, nor is time just a sequence of events which happen: both the past and the future are simply "there", laid out as part of four-dimensional space-time, some of which we have already visited and some not yet. So, just as we are accustomed to thinking of all parts of space as existing even if we are not there to experience them, all of time (past, present and future) are also constantly in existence even if we are not able to witness them. Time does not "flow", then, it just "is". This view of time is consistent with the philosophical view of eternalism or the block universe theory of time.

According to relativity, the perception of a "now", and particularly of a "now" that moves along in time so that time appears to "flow", therefore arises purely as a result of human consciousness and the way our brains are wired, perhaps as an evolutionary tool to help us deal with the world around us, even if it does not actually reflect the reality. As Einstein himself remarked, "People like us, who believe in physics, know that the distinction between past, present, and future is only a stubbornly persistent illusion".

However, if time is a dimension, it does not appear to be the same kind of dimension as the three dimensions of space. For example, we can choose to move through space or not, but our movement through time is inevitable, and happens whether we like it or not. In fact, we do not really move though time at all, at least not in the same way as we move through space. Also, space does not have any fundamental directionality (i.e. there is no "arrow of space", other than the downward pull of gravity, which is actually variable in absolute terms, depending on where on Earth we are located, or whether we are out in space with no gravitational effects at all), whereas time clearly does.

With the General Theory of Relativity, the concept of space-time was further refined, when Einstein realized that perhaps gravity is not a field or force on top of space-time, but a feature of space-time itself. Thus, the space-time continuum is actually warped and curved by mass and energy, a warping that we think of as gravity, resulting in a dynamically curved space-time. In regions of very large masses, such as stars and black holes, space-time is bent or warped substantially by the extreme gravity of the masses, an idea often illustrated by the image of a rubber sheet distorted by the weight of a bowling ball.

Time Dilation Relativity

Time dilation is just one consequence of the Theory of Relativity and curved space-time

Also as a result of Einstein's work and his Special Theory of Relativity, we now know that rates of time actually run differently depending on relative motion, so that time effectively passes at different rates for different observers travelling at different speeds, an effect known as time dilation. Thus, two synchronized clocks will not necessarily stay synchronized if they move relative to each other. There is a related effect in the spatial dimensions, known as length contraction, whereby moving bodies are actually foreshortened in the direction of their travel.

Time dilation (as well as the associated length contraction) is negligible and all but imperceptible at everyday speeds in the world around us, although it can be, and has been, measured with very sensitive instruments. However, it becomes much more pronounced as an object's speed approaches the speed of light (known as relativistic speeds). If a spaceship could travel at, say, 99% of the speed of light, a hypothetical observer looking in would see the ship's clock moving about twice as slow as normal (i.e. coordinate time is moving twice as slow as proper time), and the astronauts inside moving around apparently in slow-motion. At 99.5% of the speed of light, the observer would see the clock moving about 10 times slower than normal. At 99.9% of the speed of light, the factor becomes about 22 times, at 99.99% 224 times, and at 99.9999% 707 times, increasing exponentially. In the largest particle accelerators currently in use we can make time slow down by 100,000 times. At the speed of light itself, were it actually possible to achieve that, time would stop completely.

Perhaps the easiest way to think of this difficult concept is that, when an object or person moves in space-time, its movement "shares" some of its spatial movement with movement in time, in the same way as some northward movement is shared with westward movement when we travel northwest. What forces this sharing of dimensions is the invariant nature of the speed of light (slightly less than 300,000km/s), which is a fundamental constant of the universe that can never be exceeded. Thus, the slowing of time at relativistic speeds occurs, in a sense, to "protect" the inviolable cosmic speed limit (the speed of light).

It should be noted that, although a spaceship travelling at close to the speed of light would take 100,000 years to reach a distant star 100,000 light years away as judged by olooko on Earth, tho aotronaut in tho opaooohip might hardly age at all as he travels across the galaxy. This characteristic of relativistic time has therefore spawned much discussion of the possibility of time.

According to Einstein, then, time is relative to the observer, and more specifically to the motion of that observer. This is not to say that time is in some way capricious or random in nature – it is still governed by the laws of physics and entirely predictable in its manifestations, it is just not absolute and universal as Newton thought and things are not quite as simple and straightforward as he had believed.

Some commentators, like the Christian philosopher William Lane Craig, have suggested that there may be a need to distinguish between the reality of time and our measurement of time: according to this line of thinking (which, it should be mentioned, is not a mainstream position in physics), time itself MAY be absolute, but the way we measure it must be relativistic.

One casualty of the Theory of Relativity is the notion of simultaneity, the property of two events happening at the same time in a particular frame of reference. According to relativistic physics, simultaneity is NOT an absolute property between events, as had always been taken for granted up to that point. Thus, what is simultaneous in one frame of reference will not necessarily be simultaneous in another. For objects moving at normal everyday speeds, the effect is small and can generally be ignored (so that simultaneity CAN normally be treated as an absolute property); but when objects approach relativistic speeds (close to the speed of light) with respect to one another, such intuitive relationships can no longer be assumed.

Gravitational Time Dilation

When Einstein extended his Special Theory of Relativity to his General Theory, it became apparent that a similar time dilation effect would also occur in the presence of intense gravity, an effect usually referred to as gravitational time dilation. It is almost as if gravity is somehow pulling or dragging on time, slowing its passage. The closer an object is to another object, the stronger the pull of gravity between them (according to an inverse-square law first identified by Sir Isaac Newton), and thus the more the time drag.

Again, these effects are negligible at the kinds of gravitational differences experienced in everyday life: even though, technically, a person living in a ground floor apartment ages slower than their twin who lives in a top floor apartment of the same building (due to the difference in gravity they experience), the effect might amount to maybe a microsecond over a full lifetime. There is, however, one aspect of modern everyday life where we do experience the effects of gravitational time dilation: it has a noticeable impact on the Global Positioning System (GPS), which many of us now rely on for navigation. The orbiting satellites used by the GPS system experience significantly less gravity than the Earth's surface, and are also moving very fast, so that the time distortion effects of about 38 microseconds a day have to be specifically factored in or GPS would very quickly begin to accumulate errors.

But, just as with a spaceship travelling at near the speed of light, in the extreme gravity at the edges of a black hole, for example, substantial time differences can become apparent. A black hole spins at close to the speed of light, dragging anything in the vicinity around with it, and the huge gravitational pull of a black hole can bend and warp space-time to a substantial degree. Over the "event horizon" of a black hole – the gravitational point of no return – a hypothetical clock on a spaceship (and indeed the progress of the spaceship itself) would appear from the outside to stop completely due to the infinite time dilation effect. At the gravitational singularity at the center of a black hole, gravity and density is infinite, and all the normal rules of physics just break down. Time effectively stops, just as there is no time beyond the singularity of the Big Bang

Twins Paradox

The dilation of time also gives rise to the so-called "twins paradox" or "clock paradox", whereby a hypothetical astronaut returns from a near-light speed voyage in space to find his stay-at-home twin many years older than him (as travelling at relativistic high speeds has allowed the astronaut to experience only, say, one year of time, while ten years have elapsed on Earth). Because of the time dilation effect, a clock in the spaceship literally registers a shorter duration for the journey than the clock in mission control on Earth.

The real paradox, though, as Einstein explained it, arises from the fact that (because there is no "preferred" frame of reference in relativity) we could just as easily consider the traveler in the spaceship as the one remaining at rest, while the Earth shoots off and back at close to the speed of light. In that scenario, Einstein argued, one would expect the astronaut to age much more than the inhabitants of the Earth. In fact the "paradox" is explained by Mach's Principle: the spaceship is accelerating away at near-light speed from the bulk of the universe, whereas the Earth is not. Hence, it is the spaceship (and its astronaut) that experiences the relativistic time dilation, not the Earth.

The Real Nature of Time:
An Analysis of Physics, Prophecy, and Time Travel Experiences

3.0 Quantum Mechanics

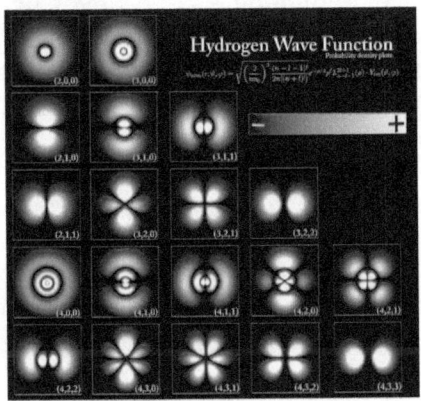

The Hydrogen Wave Function in the above composite picture.

Quantum mechanics (QM; also known as quantum physics, quantum theory, the wave mechanical model and matrix mechanics), part of quantum field theory, is a fundamental theory in physics. It describes physical properties of nature on an atomic scale.

Classical physics, the description of physics that existed before the theory of relativity and quantum mechanics, describes many aspects of nature at an ordinary (macroscopic) scale, while quantum mechanics explains the aspects of nature at small (atomic and subatomic) scales, for which classical mechanics is insufficient.

Most theories in classical physics can be derived from quantum mechanics as an approximation valid at large (macroscopic) scale. Quantum mechanics differs from classical physics in that energy, momentum, angular momentum, and other quantities of a bound system are restricted to discrete values (quantization), objects have characteristics of both particles and waves (wave-particle duality), and there are limits to how accurately the value of a physical quantity can be predicted prior to its measurement, given a complete set of initial conditions (the uncertainty principle).

Quantum mechanics arose gradually, from theories to explain observations which could not be reconciled with classical physics, such as Max Planck's solution in 1900 to the black-body radiation problem, and the correspondence between energy and frequency in Albert Einstein's 1905 paper which explained the photoelectric effect. Early quantum theory was profoundly re-conceived in the mid-1920s by Erwin Schrödinger, Werner Heisenberg, Max Born and others. The modern theory is formulated in various specially developed mathematical formalisms. In one of them, a mathematical function, the wave function, provides information about the probability amplitude of energy, momentum, and other physical properties of a particle.

4.0 Quantum Entanglement

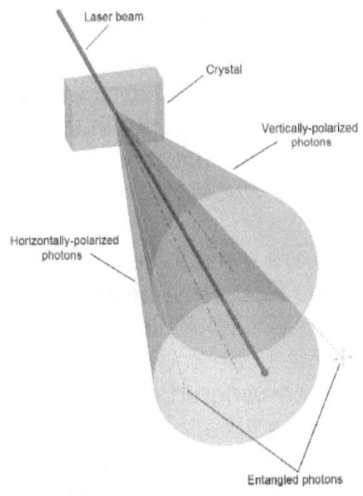

Quantum entanglement is a physical phenomenon that occurs when a pair or group of particles is generated, interact, or share spatial proximity in a way such that the quantum state of each particle of the pair or group cannot be described independently of the state of the others, including when the particles are separated by a large distance. The topic of quantum entanglement is at the heart of the disparity between classical and quantum physics: entanglement is a primary feature of quantum mechanics lacking in classical mechanics.

Measurements of physical properties such as position, momentum, spin, and polarization performed on entangled particles can, in some cases, be found to be perfectly correlated. For example, if a pair of entangled particles is generated such that their total spin is known to be zero, and one particle is found to have clockwise spin on a first axis, then the spin of the other particle, measured on the same axis, will be found to be counterclockwise. However, this behavior gives rise to seemingly paradoxical effects: any measurement of a property of a particle results in an irreversible wave function collapse of that particle and will change the original quantum state. In the case of entangled particles, such a measurement will affect the entangled system as a whole.

Such phenomena were the subject of a 1935 paper by Albert Einstein, Boris Podolsky, and Nathan Rosen, and several papers by Erwin Schrödinger shortly thereafter, describing what came to be known as the EPR paradox. Einstein and others considered such behavior to be impossible, as it violated the local realism view of causality (Einstein referring to it as "spooky action at a distance") and argued that the accepted formulation of quantum mechanics must therefore be incomplete.

Later, however, the counterintuitive predictions of quantum mechanics were verified experimentally in tests in which polarization or spin of entangled particles were measured at separate locations, statistically violating Bell's inequality. In earlier tests, it couldn't be absolutely ruled out that the test result at one point could have been subtly transmitted to the remote point, affecting the outcome at the second location. However, so-called "loophole-free" Bell tests have been performed in which the locations were separated such that communications at the speed of light would have taken longer—in one case 10,000 times longer—than the interval between the measurements.

According to some interpretations of quantum mechanics, the effect of one measurement occurs instantly. Other interpretations which don't recognize wave function collapse dispute that there is any "effect" at all. However, all interpretations agree that entanglement produces correlation between the measurements and that the mutual information between the entangled particles can be exploited, but that any transmission of information at faster-than-light speeds is impossible.

Quantum entanglement has been demonstrated experimentally with photons, neutrinos, electrons, molecules as large as Bucky balls, and even small diamonds. The utilization of entanglement in communication, computation and quantum radar is a very active area of research and development.

The Real Nature of Time:
An Analysis of Physics, Prophecy, and Time Travel
Experiences

5.0   My Spiritual Views about Reality

Although I grew up going to Methodist and Presbyterian churches I wanted to do more Spiritual exploration as I became a teenager. This exploration led to me to learning meditation and becoming more oriented towards the concepts of enlightenment. So when I started writing I explained my current beliefs about reality as follows:

I'm sure there are many paths which all go to enlightenment. Although the growth of paranormal abilities is a side benefit, In fact, the goal of spiritual growth should be enlightenment.

5.1 The Importance of Stillness

How does spiritual growth help one stay healthy; and what is stillness?

The ideas I'm going to discuss here relate to eastern Asian concepts of the spirit as taught mainly in China and India. Buddhism, Taoism, Zen, and other eastern religions and philosophies all teach that the spirit is the core of our being; and that our physical bodies are just an extension of that spirit into the physical level of existence.

By learning to let your mind or ego release its hold on the illusion of our current existence, we become aware of the spirit behind or at the core of our being. This spirit is the pure oneness of God and exists in no time and no space. (A concept which we really can't envision with our minds or egos only).

There are many techniques taught to get closer to realizing the core of a person's being. These techniques all involve practicing spiritual growth, love, and/or meditation with a goal of enlightenment.

There are thousands of books and practices on this subject so I will not try to duplicate them in this short synopsis.

The Chinese stress that the stillness and oneness obtained through spiritual growth are one of the main keys to keeping the body healthy for a long life. Many Taoist techniques and teachings stress the achievement of "stillness" as a prelude to physical immortality.
The stillness I'm referring to is found mainly through meditation. In Christian terms it is often referred to as the "Peace that passes all understanding".

It is hard to describe the feeling of stillness since it is like when you first wake up in the morning after a deep sleep—but even quieter and deeper.

The feeling of stillness has a strong effect on your body—it seems to make the randomness of your cells quiet down into a more restful state.

Meditation is taught many places, I've even found a company called Holosync which sells CDs that help even beginners achieve deep states of relaxation that usually takes advanced Yogis years of practice.

Stillness is not something achieved overnight but takes years, (even with modern advanced CD techniques) to start showing results.
However, the effects of stillness practices probably have the most profound effects on your body's aging as anything else recommended in this book.

This is since as you start to achieve stillness, your Ego is realizing its core is really part of the spirit—not a separate mind. The spirit exists outside time and space. This connection with your spirit has a profound health effect on the body in terms of peace and well-being.

When meditating in this state you can feel stillness penetrating your body. It feels like your body is reaching a relaxed state never realized; even in sleep. The state of the stillness of your spirit provides a modified blueprint for your body's health.

It is a lot of work to set aside time every day to meditate. The good news is you will find that after some weeks of practicing, this time becomes something you look forward to. This is since meditation is so relaxing it becomes a way to recharge you for daily activities in the world.

I also find that meditation makes my mind more alert when I wake up in the morning and gives me a sharper intellectual edge at work.

5.2 The Reality of Stillness

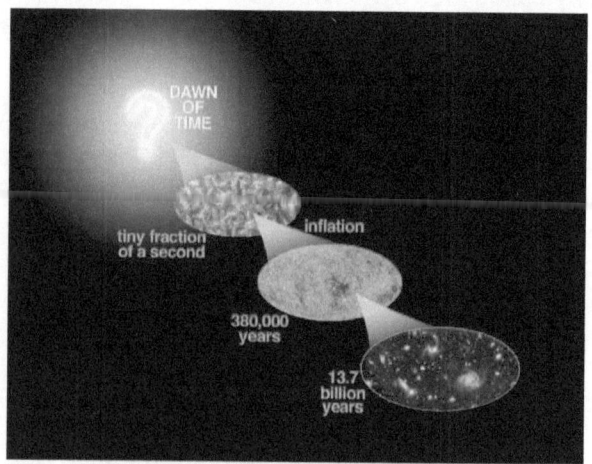

The Growth of the Universe

Most people believe that God was the initial creative force which started the Universe.

Physicists and Astronomers all agree that the Universe we know was created from nothing and inflated in a huge explosion called the "Big Bang". As it inflated time and space as we know them came into existence. When you study Einstein's Relativistic physics you being to understand that time and space are inextricably linked. You can't have one without the other.

Given our understanding of physics, we know that time and space didn't exist before the Big Bang. The state of things before creation then was "No Time & No Space".

A notional picture of a black hole

Another subject of great interest to astrophysicists is what are called "Black Holes". Black Holes are a result of Einstein's equations and astronomers have verified their existence in the last few decades. Black Holes are stars which due to their own mass have collapsed down to an infinitely small point and where time stops. Scientists do not understand where all that mass goes.

Hmm…. A Black Hole seems to be another example of part of reality that exists without time and space.
In Quantum Physics, time is also viewed differently than we perceive it on a daily basis. Here is a quote from a Physics website explaining this view:
(1)
The upshot is that, on the microscopic level, there is no direction to time -- and this is even more spectacularly true in quantum physics than in classical physics. In the microscopic domain, everything just exists in a kind of nebulous, atemporal continuum. Then, every once in a while, something becomes observable, and enters the

one-dimensional time continuum. The arrow of time does not exist in the universe as a whole. It only exists in individual subjective views of the universe!

I think it is fair to say that the place of stillness where time and space don't exist is part of our reality.

Therefore, it shouldn't be considered too strange that our immortal spirit is part of and one with that stillness.

<u>Finding Stillness in Major Religions</u>

Christianity is the largest religion in the world, and one I know pretty well since I was raised in Methodist and Presbyterian churches growing up. I also attended multiple churches as an adult and participated in Bible study groups for a number of years.

Prayer is the key to stillness as a Christian. There are many books on Prayer and Prayer techniques. One needs to focus on spirit and becoming one with the spirit to move towards a state of stillness as a Christian.

The fact that so many of the long lived persons recorded in this book lived in Christian cultures probably indicates that being a devoted Christian can help you "live in the spirit" as much as many other spiritual techniques. I'm not as familiar with Judaism and Islam, but the same approach applies in doing prayers in those religions.

The key to Prayer in your religion or spiritual approach is that you must learn to focus on the spirit of God which is inside you; and is the core of your being. That spirit exists in eternal peace; outside of time and space.

Once you learn to focus on that spirit in your prayer you will be able to bring that peace and stillness into your physical body to calm it and provide more health.

5.3 Biblical Quotes Relating to Stillness and the Spirit

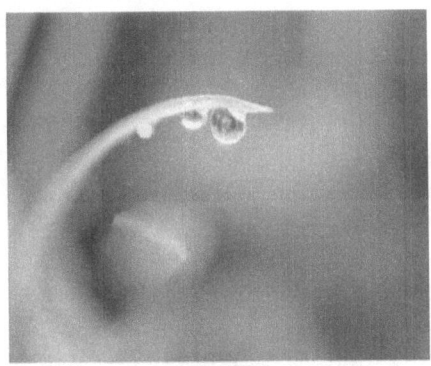

Morning Stillness

Here are a number of Biblical quotes which relate to the power of the spirit and the state of stillness (peace) I've described above.

You will experience God's peace which is far more wonderful than the human mind can understand. His peace will keep your thoughts and your hearts quiet and at rest as you trust in Jesus Christ. (Philippians 4:7 LB)
He will keep in perfect peace all those who trust in him, whose thoughts turn often to the Lord. (Isaiah 26:3 LB)
The work of righteousness shall be peace; and the effect of righteousness, quietness and assurance forever. (Isaiah 32:17 KJV)

"You shall receive power when the Holy Spirit has come upon you; and you shall be my witnesses both in

Jerusalem, and in all Judea and Samaria, and even to the remotest part of the Earth. (Acts 1:8 NASB)

For by one Spirit are we all baptized into one body (1 Corinthians 12:13 KJV)

I will ask the Father and he will give you another Comfortor, and he will never leave you. He is the Holy Spirit. The spirit who leads into all truth. The world at large cannot receive him; for it isn't looking for him and doesn't recognize him. But you do, for he lives with you now, and some day shall be in you. (John 14:16,17 LB)

Do you not know that you are a temple of God and that the spirit of God dwells within you? (1 Corinthians 3:16 NASB)

There are many more quotes about the spirit of God, but the key is that they all relate to that core of God's spirit inside us all.

6.0 Prophecy and Premonitions

I've had a lot of premonitions about the future which came true. This has led me to some interesting conclusions I'll share at the end of this chapter. Here are some of the events I've experienced:

These examples are also intended to show the reader that I do know what I'm talking about when it comes to having "sensed" the future.

6.1 Visions

During the summer of 1975 I had a summer CO-OP job at General Electric's Gas Turbine engineering group in Schenectady, NY

At this time I used to meditate at my desk during the lunch hour.

One day in early August I was meditating and thinking about a trip I was planning to Cape Cod. My mind was wandering as I was thinking about what I would do there. My thoughts went to what I would do at the beach.

All of a sudden, I had a blinding flash of a scene where I was in the surf at the beach, and a surfboard was coming towards me. Then a shock occurred and I was thrown out of my meditation and was wide-awake.

I thought that this was pretty weird, and mentioned this to a friend or two.

Two weeks later I was walking on the beach on Cape Cod. I saw a couple of guys with surfboards and asked where I could rent one to give it a try.

They said they had an extra one and I could try it with them. (I had totally forgotten my meditation vision at this point)

I tried to get up on that board all day, and had some modest success, but I was also getting exhausted in the process.

I decided to try it again and fell off when a big wave hit me. Next thing I knew I was coming up to the surface and I saw the exact same scene from my meditation.

The board hit me hard in the chin and almost knocked me out. I staggered to the shore and the two guys I was with helped me to the hospital where they put 10 stitches and 2 sutures into my chin.

The question arises—Would I have been able to avoid the accident if I had remembered my vision and not gone surfing?

Later experiences have convinced me that the future is a set of probabilities, and we have free will to decide our actions.

I also had an experience on that trip of being able to partially heal my wounds very quickly through a deep meditation and application of psychic healing techniques. However, I do still have a small scar on my chin from this accident.

6.2 Warnings of Danger

- Detroit

In 1980 I was moving from Dekalb, Illinois to Rochester, NY between assignments at General Electric, Inc.

While staying with my cousin outside Detroit, I made arrangements one evening to meet an old RPI friend Steve. We decided to go into downtown Detroit to the newly completed Renaissance Center to eat dinner and look around.

The Renaissance Center was built near the water and surrounded by slums.

After dinner we were walking out through the lower level in an area that was all boarded up with nobody else there.

Suddenly, I had this strong urge to turn around and go to find a restroom. I stopped walking forward because the urge was so strong.

I tried to walk forward again and again a very strong urge came to turn around and go back into the main center where other people were. I remarked to Steve that I couldn't go forward—that something wouldn't let me.

Just then two black guys in trench coats appeared about 30 feet away from behind one of the foundation pillars we were about to walk past. They started walking towards us with smiles pasted on their faces.

My friend Steve took off running back into the main area and after a moment or two I figured I didn't know what these guys were carrying under their coats, so I ran too.

In less than 30 seconds we were back in a populated area with Police present, and the two guys chasing us gave us smiles like "next time we'll get you" and took off going the other way.

I had previously always tried to pray to God for protection, and tried to give a subconscious message to my senses to warn me of danger.

I'm convinced that whatever sense or "angel" warned me that evening, I would have been killed or severely wounded if I had continued walking out of the complex with no warning.

- Planning a Trip to Spain

During early August of 1998, my wife and I decided to
send her and our kids to visit her mother in Barcelona,
Spain.

I was going to buy a ticket separately, and meet them there
during early September.

When I started to call the travel agent to book my ticket I
had a terrible feeling of fear about taking the flight.

I tried two other times to book the ticket during the week for
a September 2nd departure, and each time I got the same
strong feelings of fear and death.

I have always prayed and tried to guard myself mentally to avoid disasters, so finally I took the warning seriously and decided not to go at all.

This was very difficult to do since I really wanted to see my wife and kids, and this meant I would be home alone for a month.

Work wasn't an excuse either, since I wasn't doing any really heavy contract work at the time and could easily have taken the time off.

I called my wife and told her my decision, and she was surprised, but agreed for me to follow my instincts.

On September 2nd the Swissair disaster occurred on a plane leaving Kennedy airport in New York, which crashed in Newfoundland Canada with all lives lost.

I would not have originally been booked on that flight, but could have easily ended up on it since I was due to fly through Kennedy airport, and any delay might have caused me to switch planes.

I will never know for sure, but this was a very strong warning.

I should also mention that for several years before this event I had strong feelings that my I would be killed in the near future. After this happened those feelings ended.

6.3 A few Seconds Ahead

Sometimes just having sensitivity about what will happen a few seconds into the future will have a positive effect.

Avoiding a car accident at an intersection is one result.

I believe animals have spiritual abilities too.

Here is an example concerning our last dog Apollo. Some years ago he was watching my wife as she planned to start disconnecting a motor inside our dishwasher.

Apollo started barking madly at us (which he never did) and we suddenly realized that we hadn't turned off the power to the dishwasher.

He seemed to be sensing a future event which made him really worried.

6.4 Dreams of Indian Ocean Tsunami in 2004

Back around the year 2000 through 2004 I was having a series of dreams which were similar but all slightly different.

I seemed to be in a tropical coastal area and at some type of resort. There were lots of people on the beach and there were different types of resorts in each dream.

I had a lot of fear and then it happened. There would be some type of huge wave which crashed over us, or the tide would go out and a huge wave would come in and I would be covered by the wave.

When that happened I never escaped but seemed to be one of the victims.

I recall this type of dream happening at least five to ten times over that multi year period.

Of course the disaster finally happened—the late December 2004 Tsunami of the Indian Ocean which killed at least one quarter million persons.

Scientists researching this event now think this may have been the worst Tsunami disaster in over 600 years in the Indian Ocean.

6.5 Experiences about 9/11/2001

There are many reported experiences of premonitions about the major disaster of terrorists striking the two towers on 9/11/2001. Here are a couple of my own experiences: In 1976 during the summer, I had spent the last couple of years going through a spiritual and psychic development class on the side while going to college. My parents lived in New Jersey and we used to go into Manhattan once in a while. This was while the Port Authority towers were still under construction.

I was very interested in them and decided to go to the site to see how far up I could go in the towers. At the time half the building was finished and the public could go up in the elevators. I was able to get to about the fiftieth floor. I entered the floor and it was empty except for the main columns. Going over the window I wondered how long these towers would last ? I decided to try my intuition to get an answer. So I put my hands on one of the pillars next to the windows, closed my eyes and asked "How long will this building last?" I figured it would be at least one hundred years. Imagine my shock when my intuition said to me "Twenty Five years". This didn't make any sense at the time so I figured I must be wrong.

In retrospect this experience showed me that our prophecy abilities can tell us about events many years in the future.

Then in the year 2000 in September I was in New York City again visiting my sister who lived there. I was married at the time and my ex-wife had some intuitional abilities. We were pushing a stroller with our baby son in it along with my sister who had her baby daughter in her stroller. The location was in the financial district next to the twin towers. Suddenly my ex-wife said "I just got a strong impression that many people are going to die here." This is all the detail I remember. We soon forgot the incident and went on to enjoy our visit with my sister.

Some events are so large that they affect the past and many people even in different times around them.

7.0 Reported Experiences of Time Travel

Here are just a few of the experiences I collected in my book "Real Time Travel Stories from a Psychic Engineer". These stories are meant to illustrate examples of how people move through time.

Note that these stories include what may have happened by persons consciousness seeing the past, but also might have been by people passing through some type of "time warp".

7.1 Visiting Versailles in the Past

Here is a famous story of two women in 1901 who had a visit to Versailles which -shifted to the year 1792

Moberly and Jourdain recounted that they had decided to visit the Palace of Versailles as part of several trips around Paris, detailing how, on 10 August 1901, they travelled by train to Versailles. They remembered not thinking much of the palace after touring it, so they said they decided to walk through the gardens to the *Petit Trianon* but after reaching the *Grand Trianon* found it was closed to the public.

They recollected traveling with a Baedeker guidebook, but said they became lost after missing the turn for the main avenue, *Allée des Deux Trianons* and entered a lane, where they bypassed their destination. Moberly reported that she noticed a woman shaking a white cloth out of a window while Jourdain recalled noticing an old deserted farmhouse, outside of which was an old plough.

At this point they described a feeling of oppression and dreariness coming over them after which men who they thought looked like palace gardeners told them to go straight on. Moberly described the men as "very dignified officials, dressed in long greyish green coats with small three-cornered hats." Jourdain recalled that she noticed a cottage with a woman holding out a jug to a girl in the doorway, describing it as a "*tableau vivant*", a living picture, much like Madame Tussauds waxworks.

Moberly did not observe the cottage, but remembered that she felt the atmosphere change. She wrote: "Everything suddenly looked unnatural, therefore unpleasant; even the trees seemed to become flat and lifeless, like wood worked in tapestry. There were no effects of light and shade, and no wind stirred the trees."

The Comte de Vaudreuil was later suggested as a candidate for the man with the marked face allegedly seen by Moberly and Jourdain.

They reported reaching the edge of a wood, close to the *Temple de l'Amour*, and coming across a man seated beside a garden kiosk, wearing a cloak and large shady hat. According to Moberly, his appearance was "most repulsive... its expression odious. His complexion was dark and rough." Jourdain noted "The man slowly turned his face, which was marked by smallpox; his complexion was very dark. The expression was evil and yet unseeing, and though I did not feel that he was looking particularly at us, I felt a repugnance to going past him. They said that another man whom they described as "tall... with large dark eyes, and crisp curling black hair under a large sombrero hat" came up to them, and showed them the way to the *Petit Trianon*.

Moberly said she noticed a lady sketching on the grass who looked at them after they crossed a bridge to reach the gardens in front of the palace. She later described the lady as wearing a light summer dress and a shady white hat with lots of fair hair. Moberly reported that she thought she was a tourist at first, but the dress appeared to be old-fashioned. Moberly came to believe that the lady was Marie Antoinette. Jourdain, however, did not see the lady.

At their return to the palace, they reported that they were directed round to the entrance and joined a party of other visitors. They said that after they toured the house, they had tea at the *Hotel des Reservoirs* before returning to Jourdain's apartment.

The Real Nature of Time:
An Analysis of Physics, Prophecy, and Time Travel
Experiences

7.2 Bold Street, Liverpool, England

Bold Street is a site with the most experiences of time slips so far discovered. The below stories indicate that there is some type of time warp on this street which some people pass through.

The Liverpool Time Slips and Mysterious Occurrences in Bold Street are numerous. This location seems to be some type of time portal. Several stories follow.

The Bold Street Timeslips Liverpool parascience.org.uk

The subject of time has always intrigued us. Is it as set as we have always believed? Or does time loop back on itself, giving us a glimpse of a shadowy past out of the corner of our eye.

Was is just our imagination that made us believe we had seen an object or building change before our very eyes, and seem as though we had stepped back into the past? When this happens we usually shake our heads and put it down to imagination.

But over the last few decades, something strange has been happening in or near Bold Street, Liverpool England Not just a glimpse of the past, but full immersion into the strange and mysterious world of English History, if only for a few moments at a time.

The strange thing about the Bold street time slips is the actual time and place they are set. In the following cases, the people involved do not go back really far, but seem to visit a particular decade or decades.

So far, most of the sightings have centered around the 1950s and '60s. This is strange in itself. Most time travel experiences seem to take the recipient back to the 18th or 19th century. But not in this case.

Are these people simply copying each other in their experiences, or are they genuinely taking a step back in time?

The answer to this has to take into account whether they are doing it deliberately to get noticed. In other words are the perpetuating a hoax?

Another explanation could be mass hallucination.
And last but not least, they really are experiencing this strange phenomena!

The most important point is, the very first person that had this experience, obviously totally believed in what he saw, heard and felt.

So, does time flow like a river? Or does it twist and turn, going forward then sweeping back, picking up historic events and placing them down in front of you, if only for a few moments?

In this first tale, we find Frank and his wife out for a stroll in Liverpool town center. It is 1996.

His wife decided that she wanted to go and buy a book at Waterstone's the large book store, and they started to walk towards the area of the shop.

As they approached Bold Street, Frank decided to go to another shop first, but bumped into his friend, and stopped to chat in the street. His wife went ahead without him.

A few moments later, Frank said goodbye, visited his shop and turned to go back to meet his wife. After reaching Bold Street, he headed on towards the bookstore. As he approached, he glanced up and was surprised to see the name, Cripps above the door. As he was about to cross over to see what was going on, a van swept past him with the name Cardin's on the side. The van driver honked his old fashioned horn and drove past.

Looking around, Frank suddenly realized that things were not quite what they should be. He looked at the cars driving past and realized that they were all old fashioned vehicles such as people would drive back in the 50's and 60's.

And then he noticed the people. Men were wearing hats and macs, and the women were dressed in head scarves, full skirts and had old fashioned hair styles such as women wore just after the war.

By this time, Frank was beginning to feel slightly freaked out. He carried on crossing the road and headed towards the store.

As he got closer he noticed in the window there were handbags, shoes, and umbrellas. Suddenly he saw a young woman looking up at the shop sign. She looked confused.

She was wearing modern clothes and as she saw him approaching, she smiled at him.

Frank went into the shop, closely followed by the young woman. When they entered he was surprised and pleased to see that it had indeed turned back into a bookshop. The young woman smiled, shook her head and said, 'that was strange, I thought it was a new clothes shop!' then she walked away looking extremely puzzled.

This may sound an unlikely tale, but the odd thing about it is that Frank was, in fact, a former Police officer who was used to dealing in facts, and definitely wasn't the type of person who would believe in the paranormal.

Frank never stopped talking about it. Was this a time slip? Evidently Cripps was a women's shop that sold clothes and other goods decades before!

And Cardin's was also a well-known Liverpool firm that owned vans around the same time.

The second story concerns a young girl by the name of Imogen. She had decided to go into Liverpool to buy her sister Abigail a few things for her new baby. Upon arriving she was happy to see a new MotherCare store that had opened up on the corner of Lord Street and Whitechapel.

She wandered around the store, and picked up a few baby items such as cardigans, baby bibs, and gloves. She was surprised to see how cheap the items were, but thought they were on offer as the store had just opened. Taking them to the counter, she tried to pay with her credit card. The staff member looked at her suspiciously, and went off to get the manager.

When she came back, she looked at the card and told Imogen that they didn't take cards. So, disappointed, Imogen went and put the items back as she hadn't any money with her.

When she got home, she told her mother what had happened. Her mother was surprised and really puzzled. 'That store closed years ago,' she said. 'There is a bank there now, in fact that's where I have my account'. Not believing her, Imogen took her mother back to the same place the next day. Sure enough the store wasn't there. It was a bank, just as her mother had told her.

The third tale is of a young man named Sean, who, while shoplifting in Liverpool back in 2006, ran away from a Security Guard and headed down Hanover Street. Trying to shake off the Guard, Sean, 19, turned into a dead end street called Brookes Alley.

By this time he was out of breath and started to get a tight sensation in his chest. He soon realized that actually it wasn't a problem with him, but the atmosphere around him.

He waited for the Guard to come around the corner after him, but he never appeared. So, thinking he had given him the slip, he sauntered back out and started to walk down Hanover Street again. But he soon realized that something was wrong.

The road looked different, and so did the pavement. He noticed cars driving by that looked very old fashioned, and the road works that he knew were there, were now gone. Soon he saw that the people around him were wearing strange clothes. Crossing over to Bold Street, he noticed that there were traffic lights where they weren't before, and bushes growing around the Lyceum, near a bar that he recognized.

He carried on walking. Soon he began to feel that something was not quite right. Then he began to panic. He realized that somehow he had stepped back in time. And the time slip was not going away.

Then he remember his Cell phone. Taking it out of his pocket, he tried to get a signal, but of course it didn't work. Eventually he began to really panic, but soon spotted a kiosk selling newspapers and headed over.

Leaning over the Stand, he took a look at the front page of the Daily Post. There in bold lettering was the date. 18th May 1967.

He wondered what to do. What happens if he can't get back to his own time? What about family and friends?

So, speeding up his pace, he reached H. Samuel the Jewelers, and tried his phone once again. This time it worked. Sighing with relief he looked around and realized that he had returned to the present. But the strange thing was, he could still see, down the end of the road, people still walking around in 1967.

By this time Sean had seen enough, and dived onto a bus to go home. When he was interviewed by the local newspaper later, he stated over four times, the exact account.

Now, you may think that Sean was making the story up to escape from the guard. But the strange tale didn't end there. When the Security Guard was interviewed, he stated that when he ran after Sean, and turned down the dead end alley after him, he said that Sean had completely disappeared!

When the newspaper checked out the facts of Sean's story, they found that everything he said was historically accurate.

These three stories are just the tip of the iceberg. There are many tales from around Liverpool that tell of time slips, ghosts and other strange phenomenon. The stories keep coming thick and fast, and of course the more tales, the more likely people will start to believe that they are all being made up, or as the saying goes, Urban Tales. So, what do you think? Real life time slips, imagination, mass hallucination or purely tales that have started out as fun but have turned into the greatest Urban Legends of all time.

My opinion is that, yes, something did happen.

Probably to the first guy, Frank who was just out shopping with his wife. The others? Maybe it was a case of mistaken roads, taking a wrong turning or just a glitch in the person's memory. By the time they get home they totally believe what happened.

Or is it true? There are so many cases concerning Bold Street, and just about anywhere else in Liverpool, that maybe, just maybe they are all living on top of the biggest time slip phenomena in the World.

7.3 Greek Trireme, Turkey 1984

Incident at Miletus. Story is told in the first person by the man who experienced it:

In 1984 a man was travelling round Turkey with a small group of 15 or so people. We had seen some amazing sites in Eastern Anatolia and along the Lycian Coast and we were making our way back to Istanbul along the Aegean coastline. As you may be aware, that part of Turkey has an amazing number of ancient Greek and Roman sites.

On one particularly hot afternoon, we arrived at the remains of the Greco-Roman city of Miletus. I knew nothing of this place but I was looking forward to a leisurely wandering around the ruins.

Miletus was like every other site we had visited up until that time. We had the whole site to ourselves. Well, that was not quite true, there were also lots of goats and sheep, the occasional tortoise and the incessant buzz of the cicadas. But that was about it.

On arrival, I really had the urge to be alone. As my then girlfriend, Jenny, and the rest of the group headed down to the Roman theatre, we walked off in the other direction. I noticed a pink dome rising out of the trees a short distance away and I went to investigate,

'On getting closer, I realized that it was a somewhat derelict mosque. Islamic architecture has always fascinated me so I decided to check it out. Inside it was very overgrown and I decided, for some odd reason, to climb up on to the roof area around the dome. After a difficult and somewhat dangerous scramble, I found myself on the roof.

The view was splendid. I could see across the site and in the distance I could see Jenny and the others sitting in some shade next to the ancient theatre. There was nobody else around. Even the goats seemed to have left me to my solitude.

Then something really strange happened. The only way I can describe it is that the air around me became electric. I felt a tingling all over me, the view I was looking at seemed to shiver like a TV screen losing its signal. Time seemed to stop. As the air calmed down again. I suddenly found I was looking out over an entirely different scene.

Moments before I had been viewing the swaying cotton fields of the Menderes Rivers round plain but now it was different. The flood plain had turned into a wide estuary *filled with water, what had been hills in the distance were now offshore islands. The water lapped in as far as the ruined theatre that I could see further down the valley. I really could not take in fully what I was seeing but it was totally vivid, I can still see it now in my mind's eye. But what happened next snapped me back to 1984. As I looked over at one of the islands -- I saw the prow of a ship appear from behind the island, not a modern craft but a type of ancient Greek galley known as a trireme.*

It had a distinctive curved shape to the front. This was the image that was to shake me out of the 'dream' and back into 'reality'. The ancient trireme and the strange vista vanished. The electric sensation ceased, to be replaced by the buzz of the cicadas that greeted me as I returned to the 'present'.

It took me a few seconds to recover, my girlfriend, Jenny, then turned up and took a photograph of her weird boyfriend sitting on the roof of a mosque (A photograph I still possess and have now posted on my Blogsite). When I came down from the heights {both actually and metaphorically) I was still in a strange state of disinclination. A state that would take me many days to fully recover from.

When I returned to the UK, I was keen to know more about the original geography of the area. I was amazed (but out really surprised) to discover that what I had seen was exactly how the coastline looked in ancient times. The Menderes River was a large fjord-like inlet from the sea and the hills present today were then, as I had 'seen' offshore islands. Boats, mostly triremes like the one I had seen, would unload right next to the theatre.

But this event had a strange coda. A few weeks later, the Sunday Times ran an article about that part of Turkey. The reporter, a person whose name escapes me, described how when he was near 'the Mosque" at Miletus he experienced a "curious timeless state'. He said nothing more about this but I wonder what he actually meant. And had he experienced some weird time slip as I had? Somewhere in our loft is the original article from 1984. It so stunned me that I kept it.

7.4 Rudolph Fentz 1951

The Fentz legend describes how one evening in mid-June 1951, at about 11:15 p.m., passersby at New York City's Times Square noticed a man of about 20 years of age, dressed in the fashion of the late 19th century. No one observed how he had arrived there, and he was disoriented and confused standing in the middle of an intersection. He was hit by a taxi and fatally injured, before people were able to intervene.

The officials at the morgue searched his body and found the following items in his pockets:

- A copper token for a beer worth 5 cents, bearing the name of a saloon, which was unknown, even to older residents of the area;
- A bill for the care of a horse and the washing of a carriage, drawn by a livery stable on Lexington Avenue that was not listed in any address book;
- About 70 dollars in old banknotes;

- Business cards with the name Rudolph Fentz and an address on Fifth Avenue;
- A letter sent to this address, in June 1876 from Philadelphia;
- A medal for coming in 3rd in a three-legged race.

None of these objects showed any signs of aging. Captain Hubert V. Rihm of the Missing Persons Department of NYPD tried using this information to identify the man. He found that the address on Fifth Avenue was part of a business; its current owner did not know Rudolph Fentz. Fentz's name was not listed in the address book, his fingerprints were not recorded anywhere, and no one had reported him missing.

Rihm continued the investigation and finally found a Rudolph Fentz Jr. in a telephone book from 1939. Rihm spoke to the residents of the apartment building at the listed address who remembered Fentz and described him as a man about 60 years who had worked nearby. After his retirement, he moved to an unknown location in 1940. Contacting the bank, Rihm was told that Fentz died five years before, but his widow was still alive but lived in Florida. Rihm contacted her and learned that her husband's father (Rudolph Fentz) had disappeared in 1876, aged 29. He had left the house for an evening walk and never returned. All efforts to locate him were in vain and no trace remained.

Captain Rihm checked the missing person's files on Rudolph Fentz in 1876. The description of his appearance, age, and clothing corresponded precisely to the appearance of the unidentified dead man from Times Square. The case was still marked unsolved. Fearing he would be held mentally incompetent, Rihm never noted the results of his investigation in the official files.

8.0 The Characteristics of Time

So let's look at what are the different characteristics of time from the different theories and phenomena in previous chapters:

8.1 Relativistic Time Effects

- Time slows down as you approach the speed of light compared to an observer at rest (time dilation)
- Time slows down as you enter a black hole compared to being outside of one (also time dilation)

8.2 Quantum Time Effects

- Quantum entangled particles affect each other with no intervening time-no matter the distance

8.3 The Spiritual Time Realm

- Our spirit exists in a timeless realm
- This timeless realm underpins the Universe we see. It has no space and no time
- Everything in the Universe is conscious on different frequencies.

8.4 Premonitional & Prophecy Experiences

- The future has many probabilities and we can see those probabilities
- You can choose alternative futures but need to take momentum into account. Changing momentum takes will power
- Our Spirit existing in "no time and no space" can see the past, present, and future and communicates it to our consciousness

8.5 Time Travel Experiences

- Our consciousness can travel into the past
- There are locations with a "time warp" or "rip in the vale" which allow people to walk into the past
- A couple of cases indicate a person can physically travel to the future

9.0 Commonalities of Time Experiences

We can say that there are quite a few instances when time can stop or doesn't exist. This includes when reaching the speed of light, Quantum Entangled particles communicating states, being inside a black hole, and each of our spirit's time experiences since our spirit exists in "no time ad no space". So time appears to be closely related to motion in our relativistic space time Universe. It is amazing how many example we can think of about likely "no time" situations.

We can also change the probabilities in our future but it can be difficult because of the momentum towards certain events happening. My avoidance of the Swissair flight disaster is a good example.

Our consciousness can also move through time. It can move to the past or view the probable futures. People also seem to be able to travel to the past—which is the most common of these types of experiences, and a few can actually travel to the future. It might be that they travel to probable futures and we learn of it because we live in that probable future. (Rudolph Fentz is a good example.) This also supports the idea of the multiverse or multiple realities existing.

My experiences with premonitions or prophecy have led me to use it in my daily life. When I'm going to meet someone new, or am going to have some type of outing I also look at a timeline in my mind of this event and feel for impressions of what will happen—good or bad or something else. My future viewing is usually dead on.

Here are some more theoretical approaches to understanding time:

<u>Mystery of time</u>

There have been suggestions to look at the concept of time as an emergent phenomenon that is a side effect of quantum entanglement. In other words, time is an entanglement phenomenon, which places all equal clock readings (of correctly prepared clocks, or of any objects usable as clocks) into the same history. This was first fully theorized by Don Page and William Wootters in 1983. The Wheeler–DeWitt equation that combines general relativity and quantum mechanics – by leaving out time altogether – was introduced in the 1960s and it was taken up again in 1983, when Page and Wootters made a solution based on quantum entanglement. Page and Wootters argued that entanglement can be used to measure time.

In 2013, at the Istituto Nazionale di Ricerca Metrologica (INRIM) in Turin, Italy, researchers performed the first experimental test of Page and Wootters' ideas. Their result has been interpreted to confirm that time is an emergent phenomenon for internal observers but absent for external observers of the universe just as the Wheeler-DeWitt equation predicts.

<u>Source for the arrow of time</u>

Physicist Seth Lloyd says that quantum uncertainty gives rise to entanglement, the putative source of the arrow of time. According to Lloyd; "The arrow of time is an arrow of increasing correlations. "The approach to entanglement would be from the perspective of the causal arrow of time, with the assumption that the cause of the measurement of one particle determines the effect of the result of the other particle's measurement.

The Real Nature of Time:
An Analysis of Physics, Prophecy, and Time Travel Experiences

10.0   Summary

I've had many interesting experiences from premonitions and lots of interesting research on reports of time travel. I'm fully convinced that premonitions are real and so are some forms of time travel.

Also, as an Engineer with some training in relativistic physics it's very interesting to see how time is treated by our scientists and theorists.

The purpose of this comparison and analysis is to see what I could learn about time. The biggest thing I found here is that there are many situations where time is not running. In other words instances where there is "no time". I also now think that time is only relevant in certain situations. These situations mainly involve movement in space in the x,y, or z directions. Non movement is often required for many of the "no time" states.

Time is still one of the most mysterious phenomenas in our world. I will continue my explorations and meditation on this subject.

Martin K. Ettington

July 2020

The Real Nature of Time:
An Analysis of Physics, Prophecy, and Time Travel
Experiences

11.0 Bibliography

1. Patanjali;s Yoga Sutras and Space Time.
https://www.speakingtree.in/blog/pys010711time-space-continuumpatanjali-yoga-sutras176sut. [Online]

2. Relativistic Time. *http://www.exactlywhatistime.com/physics-of-time/relativistic-time/.* [Online]

3. Quantum Mechanics.
https://en.wikipedia.org/wiki/Quantum_mechanics. [Online]

4. Quantum Entanglment.
https://en.wikipedia.org/wiki/Quantum_entanglement#:~:text=Quantum%20entanglement%20is%20a%20physical,including%20when%20the%20particles%20are. [Online]